DEEP SEA FISHER:
12 THINGS TO KNOW

by Samantha S. Bell

STORY LIBRARY
MORE TO EXPLORE

www.12StoryLibrary.com

12-Story Library is an imprint of Bookstaves.

Developed and produced for 12-Story Library by Focus Strategic Communications Inc.

Library of Congress Cataloging-in-Publication Data
Names: Bell, Samantha, author.
Title: Deep sea fisher : 12 things to know / by Samantha S. Bell.
Description: Mankato, Minnesota : 12-Story Library, [2022] | Series: Daring and dangerous jobs |
Includes bibliographical references and index. |
Audience: Ages 10–13 | Audience: Grades 4–6
Identifiers: LCCN 2020019254 (print) | LCCN 2020019255 (ebook) | ISBN 9781632359391 (library binding) |
ISBN 9781632359742 (paperback) | ISBN 9781645821045 (pdf)
Subjects: LCSH: Fishers—Juvenile literature. | Fisheries—Vocational guidance—Juvenile literature.
Classification: LCC SH331.9 .B45 2022 (print) | LCC SH331.9 (ebook) | DDC s639.2023 [B]—dc23
LC record available at https://lccn.loc.gov/2020019254
LC ebook record available at https://lccn.loc.gov/2020019255

Photographs ©: photomatz/Shutterstock.com, cover, 1; ShadowBird/Shutterstock.com, 4; Design Pics Inc/Alamy, 5; sljones/Shutterstock.com, 5; Discovery Channel/Everett Collection Inc./Alamy, 6; pavalena/Shutterstock.com, 7; Nature Picture Library/Alamy, 8; Isabella Pfenninger/Shutterstock.com, 8; Norman Pogson/Alamy, 9; World History Archive/Alamy, 9; Adele Heidenreich/Shutterstock.com, 10; Ron Niebrugge/Alamy, 11; Michael Greenfelder/Alamy, 11; A.J.D. Foto Ltd./Alamy, 12; Helen Cowles/Roobelles/Alamy, 13; zaferkizilkaya/Shutterstock.com, 14; Kip Evans/Alamy, 15; Andreas Altenburger/Shutterstock.com, 15; imageBROKER/Alamy, 16; LightCraft Studio/Shutterstock.com, 17; Nhobgood/CC3.0, 17; Robert McLean/Alamy, 18; Split Second Stock/Shutterstock.com, 19; Nature Picture Library/Alamy, 19; AB Forces News Collection/Alamy, 20; Amar and Isabelle Guillen – Guillen Photo LLC/Alamy, 21; Stuart Monk/Shutterstock.com, 21; Chris Cheadle/Alamy, 22; John Wollwerth/Shutterstock.com, 22; Planetpix/Alamy, 23; Helenstonhill/Stockimo/Alamy, 24; Design Pics Inc/Alamy, 25; U.S. Coast Guard, 26; Kirk Hewlett/Shutterstock.com, 26; photomatz/Shutterstock.com, 27; Igor Kardasov/Shutterstock.com, 28; RP Images/Alamy, 29; photoff/Shutterstock.com, 29

About the Cover
Commercial fisher worker cleaning equipment.

Access free, up-to-date content on this topic plus a full digital version of this book. Scan the QR code on page 31 or use your school's login at 12StoryLibrary.com.

Table of Contents

Fishers Are Prepared

Deep sea fishing trawler.

Deep sea fishers are often far from safety. Danger can come up quickly out on the ocean. The fishers have to be prepared for everything. Each crew member should have a life jacket, or personal flotation device (PFD). These help them stay afloat if they fall overboard.

There are many other types of safety gear. Some fishers use survival suits. These help keep them warm in cold water. Tracking beacons let rescue teams find people in trouble. Distress flares, lights, and flags help, too.

The best way for fishers to stay safe is by practicing. Safety drills help them know just what to do. These include learning how to deal with fire or flooding on the boat. Fishers also practice how to react if someone falls overboard. They also work on first aid skills. Crews who do regular safety drills are more likely to know how to respond when a real emergency occurs.

Safety drills are important on a deep sea fishing trawler.

$100
Approximate cost of a high-quality PFD

- The first inflatable PFD was invented in 1928 by Peter Markus.
- It was used by the US and Great Britain during WWII (1939–1945).
- It was nicknamed "Mae West" after the famous movie star.

5

Fishers Work Long Hours

Powerful spotlights help fishers see at night.

Some fishers spend long periods of time away from home. Some may even be gone six months or more. Others both live and work in their boats. Crew members often work all through the day and night.

Because of this, fishers usually suffer from fatigue. This means they are physically and mentally exhausted. Sometimes they can take short naps. But it is still not enough time to rest and recover.

Being tired can be dangerous. When fishers are tired, they can become easily frustrated.

ARCTIC OCEAN

RUSSIA

ALASKA

Location where *Destination* sank.

Bering Sea

Aleutian Arc

They don't communicate well. They are unable to stay focused on their jobs. They may take shortcuts or risks. They can make poor decisions. Many accidents occur because fishers don't get enough sleep.

THINK ABOUT IT

Think of a time when you were tired at school. How did you do in class?

27
Number of days the ship *Destination* was at sea

- In February 2017, the fishing boat capsized near Alaska. The six crew members were lost at sea.
- Investigators believed ice caused the boat to sink.
- The captain had decided to continue on in the freezing water. Fatigue probably played a part in that bad decision.

Fishers Often Sail Through Storms

Some fishers work in all types of weather. They may be having a good fishing season. They may need the extra money. They go out in rain, high winds, and fog.

But storms often cause boating accidents. The storms can create high waves. The waves can roll the boat onto its side. They can turn a boat upside down. Sometimes the boats become lost or sink. They might run aground or run into other boats.

Some don't make it and end up aground.

Linda Greenlaw was the first female swordfish boat captain in the US. In 1993, she faced the worst storm of her career. Winds reached 120 miles per hour (195 km/h). The waves were more than two stories high. The storm grew worse in the night. But Greenlaw and her ship made it safely through it.

Linda Greenlaw.

100
Height in feet (30 m) of the waves during "The Perfect Storm"

- In October 1991, three weather systems came together over the ocean.
- The combined energy of the systems created a deadly storm. The ship the *Andrea Gail* was lost.
- The story was made into a book (1997) and a movie (2000) called *The Perfect Storm*.

Deep Sea Fishing Is Big Business

Deep sea fishing is sometimes called commercial fishing. The fishers sell fish, shellfish, and other seafood for profit. Fishers provide a lot of seafood to many countries around the world.

Some fish are illegal to catch. The fishers pick out these fish and keep the others to sell. In 2018, fishing companies made $10.7 billion in sales. This amount is expected to grow in the future. People are learning about the health benefits of seafood. They are buying and eating more.

The crab boat *Northwestern* is 125 feet (38 m) long.

$25,380

Average amount fishers earn each year

- The ship's captain is often the owner.
- The captain usually pays the fishers a percentage of the catch.
- The fishers make most of their money during the summer months.

Fishing boats can range from approximately 26 feet (8 m) to over 328 feet (100 m) in length. Fishers on larger ships usually earn more. The value of the fish that are caught also affects how much the fishers earn.

Illegal shark fin harvesting in the south Atlantic.

ILLEGAL, UNREPORTED, AND UNREGULATED (IUU) FISHING

Sometimes fishers and fishing companies break the laws by IUU fishing. Some fish without a license or in a closed area. Some collect more fish than they should. Others catch fish that are prohibited, including endangered or protected species.

Fishing Boats Can Be Damaged

Fire can be a serious problem even to vessels surrounded by water.

A vessel disaster is an event that results in great harm or damage. It involves a ship and may be caused by bad weather. High winds can overturn a boat. They can push it into shallow water. A captain may make bad decisions. Two ships may collide. A fire may break out. Sometimes the fishing boats are damaged. Sometimes they sink. Vessel disasters cause more deaths among deep sea fishers than other dangers.

Most of the time, officials can figure out what went wrong. But sometimes they are unable to find an explanation. The *Miss Hailee* was a family-owned wooden fishing boat. In November 2019, the *Miss Hailee* was on a routine fishing trip near North Carolina. Suddenly, the boat overturned without warning. It quickly sank. No one knew what caused it to sink. Four people on board escaped. But one worker died.

8

Number of people rescued from the *Princess Hawaii*

- In March 2018, the ship was hundreds of miles off the coast of Hawaii.
- Two rogue waves hit the ship and sank it.
- Rogue waves are unexpected and unpredictable. They look like walls of water.

THINK ABOUT IT

Sometimes accidents can't be avoided. What are some things you do to stay safe if an accident occurs?

Fishing Can Be Harmful to the Environment

The demand for seafood and advances in technology have led to new fishing practices. Fishers are able to bring in more fish and other marine animals than before. But this can have a negative effect on the oceans.

Bluefin tuna have been severely overfished.

Catching a lot of fish can mean more money for the fishers. But they must be careful not to overfish. This occurs when too many fish are caught. If it happens regularly, the fish don't have enough time to reproduce. The fish population decreases.

Overfishing affects the fishers and their customers. There are fewer fish to catch and sell. Many fishers are working to conserve the ocean's resources. They want every species to have time to recover.

A young shark bycatch was released back into the sea.

2048
Year people may have to stop eating seafood because there won't be enough fish

- Forty percent of the world's population depends on seafood.
- The demand for seafood is expected to increase.
- But two-thirds of the world's fish are overfished.

A WASTE OF LIFE

Bycatch refers to the fish and other marine animals that the fishers don't want. These include dolphins, turtles, and birds. They are thrown back overboard. But many are already dead or dying. Billions of fish and thousands of other creatures are lost this way.

Some Fishers Are Divers

Fishers who dive without tanks need strong lungs.

Some fishers don't use nets or fishing lines. Instead, they dive to the ocean floor to find their catch. They search the ocean floor for invertebrates. Some gather clams and mussels. Others bring up sea cucumbers or sea urchins. Divers also catch crabs and lobsters.

Some divers use the hookah dive system. A compressor supplies air through a hose. The compressor is powered by an electric motor or gasoline engine. This provides the diver with an endless supply of air. But the diver must be careful. The hose may get tangled up. The compressor could malfunction.

Many states put limits on how much can be collected. For example, in 2018, officials in Southeast Alaska placed a limit on geoduck clams. Fishers could collect 1,000 pounds (450 kg) of

Geoduck giant clams are six to eight inches (15–20 cm) long.

these large clams every week. Once divers reached the limit, they had to stop until the next week. Some states also limit the number of fishers. These policies help make fishing areas less crowded.

Sea urchins are a popular food.

300
Number of sea urchin permits allowed in California

- Only fishers with permits are allowed to catch sea urchins.
- Stephanie Mutz is California's only female sea urchin diver.
- She sells some of the sea urchins to expensive restaurants.

Fishers Need the Right Equipment

Fishers find the fish using sonar. The sonar system sends out sound signals into the ocean. The signals bounce off of the fish. A computer changes the signals into images.

Fishers use many different types of equipment to catch the fish. The gear they use depends on the type of fish. Some fishers use hooks and lines. Some lines are attached to a pole. Others are dragged along behind the boat.

Many fishers use nets. Trawling nets are cone-shaped. These nets are towed along the bottom of the sea. Fishers can

capture a large number of fish this way. A gillnet is like a wall or curtain. It hangs in the water. The fish can get halfway through the net. Then their gills get caught.

Some fishers use traps made of metal or other strong materials. Animals enter the traps but can't escape. Traps are often used to catch shellfish and octopus. Some traps are stationary. Other traps can be moved. These are often called pots.

80+
Percent of fish that are caught in nets

- Some fishing equipment harms other marine animals.
- For example, trawling nets can damage coral habitats.
- Gillnets also trap sea turtles and marine mammals.

Lost fishing nets that trap fish and mammals are known as ghost fishing.

Fishers Need Special Training

People don't need a college degree to become a commercial fisher. But they do need to take special courses. In the US, these include marine safety training classes. They provide information about the survival equipment available on most fishing boats. They cover firefighting, flooding, and cold water survival skills. Students also practice safety drills.

Training requirements are different in each country. In Canada,

A US Coast Guard safety training class.

Radios play a key role in fishing safety.

5

Number of years when US fishers should retake a marine safety class

- The training may take place over two days.
- The class helps to re-sharpen safety skills.
- Participants also learn how to conduct emergency drills on their own fishing boats.

all fishers must have a training certificate in marine emergency duties. At least one person on every Canadian boat must be trained in first aid. Fishers also must have a professional radio operator certificate.

FISHING PARTNERSHIP SUPPORT SERVICES

This organization began in 1997 in Massachusetts. Today, it helps fishers and their families all around New England. It provides safety and survival training. It offers legal and financial advice. It helps fishers and their families deal with depression and grief. Its services are free to fishing families.

Fishers Can Get Injured on Deck

Big equipment can mean a higher risk of injury.

A deck is an open, flat surface on a boat that people stand on. Smaller boats have one deck. Larger boats can have two or more. Fishers work on the decks for long hours. They work in all kinds of weather. Many fishers are injured while on deck. Some deck accidents are even fatal.

Shrimp boats have huge nets and a lot of dangerous equipment.

Being a fisher is a physically demanding job. Fishers work with heavy containers filled with bait or catch. They maintain the nets, lines, and other equipment. They often suffer from sprains, pulled muscles, and back injuries. They slip, trip, and fall on the wet decks.

Sometimes fishers get caught in the equipment. Many of these accidents happen to shrimpers in the Gulf of Mexico. Shrimp boats use nets that hang over the sides. Winches pull the nets in. Loose clothing, hands, or arms can get caught in the winches.

87

Number of fatal injuries for US fishers between 2000 and 2015

- Some injuries can be prevented.
- Fishers need the right training and safety gear.
- Engineers work to solve problems with equipment.

THINK ABOUT IT

Describe a time you tried to walk across a wet floor. It may have been in your house or near a pool. What happened?

11

Women Can Be Fishers, Too

Female fisher on deck with albacore tuna.

Deep sea fishing is hard and dangerous work. For years, only men were fishers. But today, many women are fishers, too. Just like the men, they must work hard and prove themselves.

Female fisher inspects sockeye salmon catch in Alaska.

15

Percent of fishers in Alaska who were women in 2019

- Tomi Marsh began as a cook on a fishing boat. Today, she is a boat captain in Alaska.
- Her boat is a 78-foot (24-m) wooden fishing vessel named *Savage*.
- In 2010, she published a cookbook with her sister. The book shares seafood recipes and stories of her adventures.

In developing countries, women work at the jobs that come before or after the catch. For example, they prepare the nets. They sort the fish when they come in. They clean and dry them and take them to market. In the US, women are also involved in other areas of the fishing industry. Some are scientists. Others deal with the business side of fishing. They work in research and marketing.

FEMALE "FISHERMEN"

Many women who work in commercial fishing prefer to be called "fishermen." Just like male fishers, they worked hard to succeed at the job. They've battled storms. They've helped keep the boat running. They've brought in the catch. They've been cold, tired, and afraid. They feel they've earned the title.

Fishers Can Fall Overboard

Falling overboard is the second-leading cause of death among deep sea fishers. Most victims weren't wearing a personal flotation device (PFD). PFDs used to be big and hard to wear. But today, many are lightweight and more comfortable.

26 Sometimes fishers fall when they're working

Commercial fishing is dangerous work.

alone. Other times, no one sees the fall. Between 2000 and 2016, 204 fishers died from falling overboard. Over half of the falls were unwitnessed. Most of the victims were never found.

Fishers can take steps to stay safe. Rails can be made higher. Workspaces and decks can be enclosed. Fishers can wear a PFD on deck at all times. Man overboard alarms can alert other crew members.

5

Distance in miles (8 km) a fisher was dragged behind his boat in December 2018

- A fisher named Nathan Rogers fell off his boat near the UK coast.
- He was able to grab onto the nets behind his boat.
- The boat continued forward on autopilot. Rogers was rescued when it crashed into a pier.

DANGEROUS TEMPERATURES

If a person falls into cold water, they may start breathing too fast. They might breathe in water. They may become unable to swim. They can't survive in the water for long.

More Daring and Dangerous Jobs

Boat Captain

Boat captains operate vessels on the water. They are in command and supervise all activities. They must know how to do everything on the ship, including navigating and operating all equipment. Captains are also responsible for the safety of the passengers and crew.

Marine Biologist

Marine biologists study marine organisms in their natural habitats. They may work in a lab or at a university. Some spend a lot of time in the ocean doing research. They work on ships to locate, tag, and monitor marine animals.

Marine Safety Coast Guard Officer

Marine safety officers help people who are in trouble in the ocean. They investigate boating accidents. They conduct search and rescue missions. They save lives and property. They also teach people how to stay safe while boating.

Glossary

autopilot
A device that automatically pilots or steers a ship.

capsize
To accidently overturn a boat in the water.

collide
To hit or run into something forcefully while moving.

compressor
A machine used to supply air to a diver.

developing
A country that is seeking to become more advanced economically.

flare
A tube packed with explosive chemicals that burn to attract attention in an emergency.

gill
The organ a fish breathes through by taking oxygen from the water.

invertebrate
An animal such as a clam or mussel that does not have a backbone.

malfunction
Failure of a piece of equipment such as a compressor to work properly.

prohibited
Something that is forbidden, Illegal, or not allowed, such as catching certain types of fish.

winch
A device used to lift or haul something such as fish nets, using a rope, cable, or chain.

Read More

Hogan, Zeb and Kathleen Weidner. *Monster Fish!: True Stories of Adventures With Animals.* Washington, DC: National Geographic Kids, 2017.

Labrecque, Ellen. *Commercial Fisherman.* Ann Arbor, MI: Cherry Lake Publishing, 2016.

Miloszewski, Nathan. *Commercial Fishers.* New York: Rosen PowerKids Press, 2019.

Visit 12StoryLibrary.com

Scan the code or use your school's login at **12StoryLibrary.com** for recent updates about this topic and a full digital version of this book. Enjoy free access to:

- Digital ebook
- Breaking news updates
- Live content feeds
- Videos, interactive maps, and graphics
- Additional web resources

Note to educators: Visit 12StoryLibrary.com/register to sign up for free premium website access. Enjoy live content plus a full digital version of every 12-Story Library book you own for every student at your school.

Index

About the Author

Samantha S. Bell has written more than 125 nonfiction books for children. She also teaches art and creative writing to children and adults. She lives in the Carolinas with her family and too many cats.

READ MORE FROM 12-STORY LIBRARY

Every 12-Story Library Book is available in many formats. For more information, visit **12StoryLibrary.com**